GAMES AND UNIVERSES

POCKET EDITION

Published from
Mardukite Borsippa HQ, San Luis Valley, Colorado
Mardukite Academy & Systemology Society
for spiritual or educational purposes only

GAMES AND UNIVERSES

Systemology
Professional Course
Booklet #12

Developed by Joshua Free
for the Systemology Society

© 2023, JOSHUA FREE

ISBN : 978-1-961509-37-5

Pocket Paperback Edition — *December 2023*

mardukite.com

Chart Your Flight For Ascension...
Then Let Your Spirit Fly!

Unlock your ultimate spiritual potential by removing barriers to your true native state.

Learn how to easily attain Self-actualization and help to actualize others along the way.

A greater appreciation and understanding of *Spiritual Life* and *Existence* awaits you. Expand your reach to achieve your dreams.

Each 'Professional Course' lesson-booklet offers simple exercises and techniques that directly apply the philosophy of Systemology, assisting to increase your true knowingness, improve your capabilities in this life, and even decide what you will do in your next.

At the Mardukite Academy of Systemology, the 'Professional Course' lessons in this series are presented to Seeker's that have already completed the 'Basic Course', previously released as six lesson-booklets, or the six-in-one single volume edition "Fundamentals of Systemology."

This all new presentation of the Systemology 'Pathway-to-Ascension' takes Seekers and continuing students from "Zero" to "Infinity" at lightning-fast speeds!

Discover Who You Really Are...

Because You Were Never Human

...more titles in this series coming soon!

TABLET OF CONTENTS

COURSE INTRODUCTION

LESSON TWELVE:
GAMES AND UNIVERSES

APPENDIX

PROFESSIONAL COURSE INTRODUCTION

WELCOME, SEEKER!
LET'S CHART YOUR JOURNEY
ON THE PATHWAY

Systemology is a "holistic" approach to understanding the human experience. It is not actually a singular "subject" in itself, but rather, a new way in which to view the many subjects of *Life* and all *Existence.*

This is a professional course in *Systemology*—specifically, how to *apply* the spiritual philosophy of *Mardukite Systemology* as a personal "*Pathway*" *to Ascension.* Our *Systemology* is a new approach to "*Self-Actualization.*" It is completely relevant for the modern age and the future; and quite different from any previous similar attempts, or other traditions, you might find. What's more: it is applicable to anyone with any background.

This *"Professional Course"* series of lessons (booklets) immediately follows the material given in the *"Basic Course"* series—available as six separate pocket-sized booklets, or in a single hardcover volume titled: *"Fundamentals of Systemology: A New Thought For The 21st Century."*

This is a *new* presentation of *Systemology*, emphasizing the application of our philosophy for those *Seekers* that are *"Flying-Solo"*—or else working through their studies and exercises as solitary practitioners. This is a new innovation for *Systemology*. Aside from the book *"Crystal Clear,"* all of our former advanced courses have placed a focus on *"Traditional Piloting"*—where experienced practitioners assist *Seekers* in *"processing."*

To receive the greatest benefit from this study: it is expected that a *Seeker* will already be familiar with the fundamental concepts and terminology (previously relayed in the *Basic Course*) before using lessons from the *Professional Course*. This will allow us to cover the extensive territory of the *Pathway* much more quickly. However, for reference, a basic *"glossary"* of vocabulary used in this lesson is provided in the *"appendix."*

A NEW VIEW OF THE HUMAN SPIRIT

Systemology is not a religion and does not require any type of *faith*. It is, however, built upon a "spiritual" premise—and as such is an "applied spiritual philosophy." It is based on ancient teachings that we are *Spiritual Beings* essentially "wearing" bodies like clothes—or using them as "vehicles." Yet our true native nature is not *physical,* but beyond this existence; and we can certainly operate a "body" from *outside* of it.

We are **all** *Spiritual Beings*—each of us a *unit* of *Spiritual Awareness*—that have experienced a very long *Spiritual Timeline* of existence. Although we might be particularly attached to the familiar "physical shells" associated with *this* lifetime, our true *"Spiritual Lifetime"* is seemingly *eternal*. We have been many things before *Human*, and we go onward as a *Spiritual Being* after our *"genetic vehicle"* of *this* incarnation perishes.

While a "spiritual" view of the *Human Condition* may not seem unique to our philosophy, just how often is the concept treated *systematically*? For that matter: just how many people, supposedly raised to this or that religion, or professing to believe one thing or another, actually live their lives as though they are *Spirits?*

As *Spiritual Beings* of immortal existence and infinite potential, we are not simply the *"creations"* of an even greater *Being-*

ness; we are, in fact, an integral part of that *"creative force"* which permeates all existence.

Our basic nature is to be a *"creative being"*—our highest goals are *"to create."* And as such a being—which we refer to as an *Alpha-Spirit* in *Systemology*—we have run into some difficulties along the course of our *Spiritual Timeline* and found ourselves trapped within material *Universes* of our own collaborative *creation*.

Since we did not start out our existence in a trapped condition, it is correct to say that we have *"fallen"* from our native *"godlike"* states. It did not happen all at one, but progressively and systematically. We know our "troubles" have resulted from accumulated "barriers" and "blockages"—or *fragmentation*—during our vast experiences as *Spiritual Beings*. They are not because we lack something; but because of what's been added.

In *Systemology*, we systematically examine those routes by which we must have descended to reach our present condition, then reverse the direction of travel and chart a personal *"Pathway to Ascension."* Of course, the exact "details" of the *Spiritual Timeline* will be different for each individual *Seeker*. However, we have been able to systematically chart our *Pathway* based on common patterns of *Human fragmentation*.

In the most basic terms: the *fragmentation* that defines our "downward spiral" consists of decisions or considerations where we deny our true nature. This includes those decisions to *"withdraw"* rather than *"reach"*; where we choose to *not-know* rather than *know*; to *not-communicate* rather than *communicate*; and ultimately, to take *no-responsibility* for being a *creative-cause*, and therefore succumb to being an *effect*.

But there is *hope!* And much more importantly: there is an effectively workable *way out* of the mazes and traps of our existence. If you are reading this now, you have already begun to gather your tools and build up the *"horsepower"* necessary to break the gravity holding your *Spiritual Beingness* to the *Human Condition.*

STUDYING THE PROFESSIONAL COURSE

Most *Seekers* study and practice *Systemology* at-a-distance and independent of the "Mardukite Academy" or any "Master-level" mentors trained therein. This means that the *books* (and to a lesser degree, the *internet*) are the only means of direct contact a *Seeker* maintains with the "Systemology Society" during their studies. A continuing *Seeker* from the *"Basic Course"* will be familiar with the style of study found in *this* course.

Misunderstood words are the most common reason an individual abandons studying a subject. When a misunderstanding occurs, *Awareness* declines. These misunderstandings start to "stack up" after the first occurrence, and as a result, the level of interest and attention will also decline. This is how a "confusion" develops; and the individual will get "bored" with the subject, feel tired, and unable to concentrate.

One solution is to return to the part of the material that was still interesting and enjoyable to read. When scanning around that area of text, there is likely to be a new word (or new specific use of a familiar word) that is unclear, but was passed by unnoticed. All *Systemology* books include their own *glossary*. Using this *glossary* and a high-quality dictionary will help resolve this misunderstanding once it is located.

An effective education of any subject is taught on a *gradient*. This is what is intended by presenting the study of something as "*grades*." Rather than treating a subject as one total mass, true learning is achieved by increasing one's understanding with a *gradual* increase upward. The *ascent* to a mountaintop is not successfully achieved in one leap, but by targeting and reaching specific checkpoints along the way.

This *Professional Course* consists of a series of lessons (booklets) that gradually increase a *Seeker's* ability to understand and apply the practices and techniques of *Systemology* as a complete "*Pathway to Ascension*." It is an appropriate study for continuing *Seekers* (from the *Basic Course*), but also "advanced" *Systemologists*.

Each lesson (booklet) of the *Professional Course* applies *Systemology* to a particular subject (or focus). It is best if the entire course can be studied and applied in sequential order. These lessons also employ a style of practice or technique called "*Systematic Processing.*" An introduction to applying this methodology is provided in the final lesson (booklet) of the *Basic Course*—or in the "*Fundamentals of Systemology*" volume.

To study the *Professional Course* just like a student at the Academy: a *Seeker* reads through all instructional material and applies each exercise (or "*process*") presented in the text to the extent they comfortably can, before continuing on to the next lesson (booklet).

When first starting on the *Pathway* as a *Solo* practitioner, without the aid of an experienced *Pilot*, a *Seeker* shouldn't "push too hard" or allow themselves to get too "stuck" on any one area (lesson) or *process*. It is not expected that any one area will be completely handled when first introduced. For optimum results, it is expected that a serious *Seeker* will make more than one "pass" through the entire *Professional Course.*

The *Professional Course* is not altogether different from other forms of practical or technical education: where the instruction and exercises are delivered to a completion, and then a student further increases their abilities, strength and skill-level by applying additional practice throughout their life. Therefore, a student should not concern themselves with perfectly mastering each step (or lesson) before progressing forward.

Additional passes through the material are likely to result in different "*realizations*" (an increased *level of understanding*) than a previous time. New "layers" of *Knowingness* may now be accessible during a *process* that may not have been before. It is important to avoid invalidating

the progress you've made just because
one area is not completely handled right
away, or if a certain *process* seems too difficult on the first pass.

CHARTING A COURSE ON
THE PATHWAY

Although we can communicate a systematic structure to *fragmentation,* the personal journey experienced along the *Pathway*
will be different for each *Seeker.* For example, certain areas will seem more "*turbulent*" or difficult for one *Seeker* than
another. We tend to say that these areas
have more "*charge*" on them—or that
they are more "*heavily charged.*" It is best
to handle such areas when you are
already feeling "good" and not in a situ-
ation (or condition) where that specific
area is consistently being "*triggered*" or
"*restimulated.*"

As an applied philosophy, *Systemology* "theory" can be easily utilized in the "laboratory" of the "world-at-large" in everyday life. This is implied within the basic instruction of each lesson. Unlike other "sciences" that conduct experiments by making a change to some "objective variable" *out there* and waiting to see an effect, our focus is the individual (or *Observer*) themselves, and how *they* affect the "*Reality*" perceived.

In addition to applying *Systemology* "New Thought" to everyday life, our philosophy is applied by using specific exercises and systematic techniques. These "*processes*" provide the most stable personal gain (and *realizations*) for each area; but only when actually applied with a *Seeker's* full "*presence*" and *Awareness.*

This *Professional Course* is designed so that it may be easily read and studied with little concern for what "dangers" these teachings—or *processing*—might unleash. However, there are still some guidelines that pertain to the "best-uses" of these course lessons, particularly if a *Seeker* intends for stable development.

Skipping over too much material/*processing* in early lessons may make attempts to understand (or apply) later lessons more difficult. However, once the complete *Professional Course* is worked through at least once in its entirety, specific areas can then be later returned to and treated with a greater sense of *Awareness* and *"presence"* than before. Of course, in *"Traditional Piloting,"* the rate of processing is monitored by an experienced practitioner; but in *"Solo-Processing,"* a *Seeker* must regulate their own progress on the *Pathway*.

Applying a systematic technique is called *"running a process."* The *processes* are designed with very simple instructions or "*command-lines.*" To *run* a *processing command-line*, a *Seeker* may be assisted by the communication of that *line* from a "*Co-Pilot*" (as in "*Traditional Piloting*"). But even then, a *Seeker* must still personally "input" the *command* as *Self*. For this reason —and quite thankfully—*Solo-Processing* is possible.

TAKING FLIGHT ON THE PATHWAY

Processing Techniques are intended to treat the *Spiritual Being* or *Alpha-Spirit*; the individual themselves. It is applied by the *Alpha-Spirit*—then *Self-directed* to the "Mind-System" or even a "body" (*genetic-vehicle*), both of which are "constructs" that the *Alpha-Spirit* (*Self*, or the "I-AM" *Awareness unit*) operates, but neither of which is actually *Self*. *Fragmentation* causes *Humans* to falsely identify *Self as* the "*Mind*" or even a "*Body*."

The *Professional Course* lessons (booklets) are designed for the *Beginning Seeker* in mind—one that may have an understanding of theory, but with little experience in practice. That being said: each of these lessons may be used toward total *Beta-Defragmentation* within a specific area. There are also more *processes* given for each subject than may be necessary to achieve an *ultimate end-point realization* on that entire area.

Some *processes* can be treated quite lightly at first; others may require a bit of working at in order to get *"running"* well. It is important to set aside a period of time when you can be dedicated to your studies and *processing.* This period of time is referred to as a *"processing session."* The reason for this, is that when a *process* does start *running* well, it is important to be able to complete it to a satisfactory *"end-point."*

The purpose of *systematic processing* is to be able to *really* "look" at things and even determine the *considerations* we have made—or attitudes we have decided—about *Reality* as a result of those experiences. It doesn't do us much good to simply "glance"—or to *restimulate* something uncomfortable and then quickly *withdraw* from it once again, leaving more of our *attention* yet again behind and held fixedly on it.

Generally speaking, a *Seeker* continues to *run* a *process* so long as something is "happening"—which is to say, the *process* is still producing a change. Usually this is evident by the type of "answers" that a *command-line* helps a *Seeker* originate from the database of their own *Mind-System*. The *command-lines* do not "do" anything on their own. They assist a *Seeker* to direct their own attention toward increasing *Awareness*.

Of course, a *Seeker* may also cease to generate new "data" from a *process* without reaching an *"ultimate" realization* as an *"end-point."* It is possible that additional "layers" (or even other "areas") require handling before anything "deeper" is accessible. If this is the case, end the *process*. But, if a *Seeker* is *withdrawing* from something uncomfortable that was incited or stirred up, then a *process* is *run* until they feel "good" about it.

In case the thought of encountering *"turbulence"* is a concern: the techniques given as *"Opening Procedures"* of a *Formal Session* (in the *Basic Course*), and those found in the earliest lessons of the *Professional Course*, are quite useful when applied as "safety nets" for maintaining *Awareness* and *presence*, even when *Flying-Solo*.

One of the benefits to *Flying-Solo* is that *processing* is entirely *Self-determined*. This already provides a certain built-in "safety" for a practitioner. Anything you *restimulate* by *Self-determinism* is *your thing*. It is not incited by external *other-determined* influences (or other "source-points" in existence) that make you an *effect*. It can be more easily handled in *processing*—or you can simply let things "cool down" and come back to it again.

While it may seem "mysterious" to beginners, a *Seeker* gets a sense for knowing how long to *run* a *process* only with practice. Once you have spent some time actually applying the *Professional Course*, there are many aspects that become "second nature" because they are, in fact, a part of our true original nature. All we have done is "*reverse engineer*" the routes of *creation* and *consideration* that are already *our own*.

LESSON TWELVE: GAMES & UNIVERSES

ENTRY INTO GAMES

In our previous lesson, we introduced *advanced incident-running* of *"spiritual implants"*—systematic suggestions from an external/outside (*other-determined*) *source* that have a tendency to affect the high-power *decisions* and *considerations* that we, ourselves, make about *"What-IS."* This brings us, then, to an upper-level point on the *Pathway* where the subjects of *"Games"* and *"Universes"* are handled more directly.

Participation in a *"game"*—whether it be the experience of this *Universe* in general, or another more specific activity involving "choices"—concerns the interaction between an individual (or individuals) and preset *"conditions."* The manner in which these *conditions* (or *aspects*) are handled (and *confronted*) is what distinguishes the experience of one *"player"* from another. It also gives some insight into what exactly an individual is attempting to accomplish—their *goals* and *purposes*—while *playing* the *"Game of Life."*

There is a *systematic sequence* (or *pattern*) to the way we are preoccupied with the *"roles"* and *"goals"* of our existence. This is quite helpful for retracing the unique steps of our *"spiritual descent"* on the *Backtrack.* But although we have uncovered *cyclic patterns* on the *Backtrack,* all individuals have not walked the same path *simultaneously* with each other. A *Seeker* is likely to be at a different *phase* of a *cycle* (or perhaps a different *cycle* altogether) than the next *Seeker*—just as those of us on the *Pathway* are all working at different paces.

Although we have tread into *advanced* territory, we will start off lightly with our *"Systemology of Games"*—first, by making certain a *Seeker* understands what is meant by "*games.*" We do not only mean a *game* like "*chess,*" although that *is* a *game.* We mean anything that involves an individual *playing a role* or *interacting with others*—especially when that *participation* involves working towards a particular *goal* or *purpose.*

Those *Seekers* with a large amount of experience with *systematic processing*—and additional *advanced ability levels*—have determined that "*Spotting*" the "*entry point*" of participation (when you "*entered*") in a *game* (or even the *decision* to *participate*) can significantly "*release*" some of the *turbulent fragmentation* one might encounter when *processing* the *incidents* associated with that *game.* It is particularly effective if you can really "*Spot*" the point when your own interests and desires (*considerations*) drew you in to the *game.*

By itself, this practice does not automatically *defragment* all of the subsequent *incidents* that are connected. However, in these *upper-levels*: we are concerned with increasing greater *Actualized Awareness* in larger areas by "unfixing" (or "freeing up") personal "*attention-units*" (personal "*spiritual energy*") still "stuck on" (or suspended in) "heavier" *incidents*. By *confronting* these *incidents* as "external" from ourselves—or by seeing ourselves as "exterior" to the *incidents*—the magnitude of effect they have on us lessens enormously.

To start with: we will *consider* our involvement with others ("*relationships*" and "*groups*") as *playing* a *game* or *role* in some way. We don't only mean "*sexual partnerships*," where *relationships* are concerned, but any intense contact or involvement.

A *Seeker* might "*scan*" (or "look over") the events of a *relationship* to see if there is any "*charge*" or "*fragmentation*" present regarding a particular *incident*; but that is not really what we are treating at this level. We are most concerned with *"Spotting"* the *"entry points"* into any *relationships*, *group participation* (or *membership*), and *games*.

One might notice that happily established *"couples"* and *"partnerships"* will often "reminisce" about their own shared *"entry-point"* into the *"relationship-game."* They frequently remind themselves (and each other) about *"how they got together"* in order to strengthen the "bonds" of the *relationship*—and to reinforce the "goals" that were (and are) "shared" for their own *"game."*

One might also notice that more *"turbulent"* and unpleasant *relationships* consist of a lot of *"fluctuation"* in *attention* on the *"upsets"* that occur, rather than *shared-goals*. Some of these *"upsets"* stem from *misconceptions* that occurred near the beginning—if not at the actual "*entry point*"—of the *relationship*.

Rather than emphasize *mistakes*, a *Seeker* should focus on "*Spotting*" the original *desires* and *goals*—and the thing that you *thought was there*—while "*processing-out*" a *past-relationship*. Again, it is the "*entry-point*" into a *game* (or any *incident*)—the beginning, when it all started—that is most critical to "*spot*" when *processing*. This also takes unnecessary "*weight*" off of the things that happened later on in the *relationship*, which in turn makes any other related *defragmentation* or *incident-running* (given in earlier lessons) much easier.

Using what you have learned from earlier lessons (and techniques), if whatever you are "*processing-out*" seems to get "*heavier*" or more "*charged up*," then you need to look for an "earlier beginning" (or *entry-point*) to that *incident* (or *game*); and if that isn't working, then look for an earlier similar *incident* (or *game*) that may be getting *restimulated* as a "*chain*." By this point, a *Seeker* will be experienced in handling these technicalities of *systematic processing*.

PARTICIPATION IN GROUPS

The next introductory area of *games* we treat is our participation in "*groups*." First, "*Scan*" the portion of the *Backtrack* (your past) that is accessible and list the various "*groups*" you *knowingly* (*willingly*) joined or were a part of (that you consider significant to your life-experience).

As described in the previous section, *run* the "*entry-point*" into each *group*. And as with "*relationships*," we are interested in the "*goals*," "*intentions*," and "*considerations*" that went into the *decisions* that led up to actually joining the *group*.

Another thing to pay special *attention* to —whether in joining *groups*, or handling *relationships* (previously)—is whether the *decision* was made as a *solution* to a previous "*problem*," or as a remedy for a period of "*confusion*." This is very important for *systematic processing*; because if that is the case, the "*decision*" to enter a *game* is actually in the "middle of the story," so to speak. Here, we would want to also "*Spot*" the *conditions* that existed prior to the "*decision*"—and of course, *what* that *decision* was attempting to *solve*. If this is case, *run*:

A. "*What problem (or upset) did you have?*"
B. "*What communication did you leave incomplete about that problem?*"

An alternative approach, assuming the "*problem*" was identified:

A. "*What did you do at that time?*"
B. "*What didn't you say at that time?*"

This should help take the *"confusion"* out of the *"chain"* — which allows a *Seeker* to handle the actual *"upsets"* or *"fragmentation"* that occurred (or was *restimulated*). The *"confusion"* part is what makes *defragmenting* the "residual effects" more difficult. The remainder is handled as described for *"Preventative Fundamentals"* in *Lesson-9*; which includes *"Flow-Factors"* (*Lesson-7*), *"Human Problems"* (*Lesson-4*), and *"Hold-Outs"* (*Lesson-6*).

There are also times when our participation in a *group* (or *relationship,* or any *game*) was *"enforced"* — in that we were *"forced"* to join or participate. Usually, our personal *protest* in such cases stems directly from the fact that we *perceive* the *"decision"* as *"other-determined"* from an external *source*, rather than *Self-Determined.* Our "power" of "free choice" has been reduced. To *"process-out"* such *fragmentation*, a *Seeker* would first *process* toward greater *"relief"* for the area of *"enforcement-to-join."*

41

A. *"Recall being forced to join something."*

B. *"What were you protesting then?"*

C. *"Recall forcing another to join something."*

D. *"What did they protest about it?"*

Once a *Seeker* has elevated their level of *Awareness* (and *ability-to-confront*) in this area, then they simply *"Spot"* any significant *entry-points* where their participation in a group was enforced, and then apply *incident-running* procedures (as learned throughout the *Professional Course*). It may be that all *"circuits"* (*you enforcing others; others enforcing others*) must be *run* fully in order to get the full *"release"* from this area.

BASIC "GOALS" AND "PURPOSES"

When an individual develops a strong *"purpose," "goal,"* or *"intention,"* it is an *entry-point* (or beginning, of sorts) into a *game*—because *goals* and *purposes*, by definition, are "*game-conditions*." In fact, a *game* is primarily distinguished by its specific *goals*: something an individual is trying to *do, have,* or *Be.* This is different from "*pure creation*," because a *game* also consists of specific "*rules*" or "*parameters*" that define certain "*barriers*" and how a "*player*" is permitted to apply *effort* toward attaining the *goal.*

Sometimes a *goal* or *purpose* goes "unfulfilled" and is eventually "abandoned" altogether due to difficulties or failures in attaining it. These leave us with *fragmentation* of a "*failed purpose*"—whereas in a state of high-power *Awareness*, we might simply "set aside" one interest while pursuing another for the time being. As with other *entry-points* (described previously), "*Spotting*" the moment of the original *decision* (or "*postulate*") greatly assists with *defragmenting* anything connected to it thereafter.

As with other areas *defragmented* for the *Pathway-to-Ascension*, our *processing* is not intended to enforce any *considerations* about what a *Seeker* ultimately decides to *do*, or *participate in*, or *have*, &tc. The purpose of *defragmentation* is to rehabilitate the original power of "*free choice*" that an *Alpha-Spirit* experienced before all of their *attention* or *Awareness* became confined to the most restrictive of *considerations* (and "*postulates*") for the level of *game* taking place in *this Physical Universe*.

Once *fragmentation* is eliminated from a "*failed purpose*," the basic desire behind the underlying *goal* may be *revitalized*, or the *Seeker* will see it "*As-It-Is*" and lay it to rest depending on what their *actual goals* are for the present. These older ones (considered "*failed*") tend to get in the way of our clear handling of present *goals*— mainly due to the amount of *attention-energy* still suspended *unknowingly* in that area. When this *attention* is freed up, you can more clearly *consider* all of the *goals* you could have.

A. "*Spot a goal (or purpose) desirable to you.*"
B. "*Spot a goal (or purpose) desirable to someone else.*"

As an additional step further—for *advanced Seekers*—on future passes through the *Professional Course*, it is very beneficial to "*Spot*" and *run* the *incident* of "picking up" your current *Body*. This is perhaps one of the most critical (and earliest) *entry-points* an individual has concerning the *Game* of *this* lifetime. Of course, we have picked up many bodies, and have had repeated *incidents* of *entering into this Universe* before; hence this area requires *systematic processing* by a very experienced practitioner.

ENTRY INTO UNIVERSES

It takes a high-level of *Actualized Awareness* to fully understand and appreciate the phenomenon we refer to as the *"Condensation of Universes."* Many esoteric and spiritual traditions have produced various mystical models of *"interdimensional trees,"* *"star-gates"* and *"kabbalahs"* to demonstrate this—but too often the true intended meaning has been lost to thousands of years of reinterpretation and cross-communication.

An *entry-point* into *this Universe* (or version of *Beta-Existence*) differs from the *entry-point* into an *earth-bound* experience of the *Human Condition*; although, as we've described (above), that too is an *entry-point*. There are obviously many levels of *"game"* taking place simultaneously—and part of *upper-level defragmentation* involves *"Spotting"* and *"differ-*

entiating" between the various *games* (and their "*parts*") that we are still participating in, *knowingly* or otherwise.

Of course, as an *Alpha-Spirit* we are essentially "*non-local,*" but as we have covered in other lessons: the *condensation* (*compactification* and *solidification*) of our total *Awareness* as *Self* has become fixed to (or "*collapsed-in*" on) one *singular viewpoint*— and in our case, most recently on the *Backtrack*, fixed to a location "*interior to*" the *Human Condition,* on *Planet Earth*, in *this Physical Universe.*

Our true *Beingness* is not *actually located* behind some "*eyes*" or in a "*head,*" but for all intents and purposes, this is how we have been accustomed to operate as a "*player*" in the more "*localized*" *game* of *Beta-Existence.* At some point on the *Backtrack*, an *Alpha-Spirit* became dissatisfied with operating a *game* while *exterior-to* it, and "*decided*" that a more substantial "*immersive first-person*" level of experience was the solution. This was later added to with a reinforcement of *"pleasure"* and *"pain"* — and to take things a step further and worse: eventually an individual *"decided"* to *forget* they were *playing a game* altogether.

Throughout our *Systemology,* we often draw a distinction about *"this"* Universe or *"this"* version of *Beta-Existence* — which just summons the mind to question what *"other"* Universes or versions of *Beta-Existence* we might otherwise be referring to. [If at any point this all seems *"too speculative"* for a *Seeker,* return to this subject on your second pass through the *Professional Course.*]

Of course, we have previously mentioned (in former lessons) the *"Personal Universe"* (or *Home Universe*) where the *actual Beingness* of the *Alpha-Spirit* still remains, and of which the experience of *Beta-Existence* is *"superimposed"* over. But *this* present version of *Beta-Existence* is not the *only Universe* that has ever been "super-imposed" and experienced, as new data from the *Backtrack* consistently reflects and confirms.

We covered a basic understanding of *implants* in the previous lesson. The *entry-point* to *this Universe* also includes an "implanting-incident" in which to "transition" an individual's *Awareness* (and sense of *Beingness*) *out of* the "previous" *Universe.* The two *Universes* are not otherwise spatially connected in any way. You could not, for example, travel any distance in *this Universe* and somehow find yourself in the former one.

The only "bridge" between *Universes* is the *"implanting-incident"* itself. And the entire *implanting-incident* is manufactured (or *created*), *existing* within its own *space-time* just as any other *Universe.* Its range of potential experience is much "smaller" relatively (compared to *this Universe*, for example), since it operates on a "*prerecor-*

ded loop" of sorts; but it is obviously quite efficient in "*implanting*" our perceived parameters for the level of *game* taking place "down here" in *this Universe*.

In order to be *systematically concise* in our work, we often draw a distinction between *Universes* by referring to *this* version of *Beta-Existence* as "*this version*" or the "*Physical Universe*." By relative contrast, the previous or former *Beta-Existence/Universe* was quite "*magical*"—and hence we distinguish it with the title "*Magic Universe*" or "*Magic Kingdom Universe*." It is not simply some other "*planet*" within *this Universe*, or even an obscure "*astral*" dimension "*in-between*" *space*, or anything like that; it is an entirely separate *Universe*.

The entire sequence concerning our *Entry-into-this-Universe* really has an "earlier beginning" in the *Magic Universe*. The *transition point from* the *Magic Universe* is often described as being "pushed in to" a surrealistic (usually outdoor) pool, backed by *Greco-Roman*-style pillars and a beautiful sky. We could outline the full sequence of events briefly as this:

1. *Magic Universe* (previous "*Beta-Existence*")
2. A *transition* point *from* the *Magic Universe*
3. A "*mini-verse*" bridge (recorded *implanted-incident* between *Universes*)
4. *Physical Universe* (entering this "*Beta-Existence*")

Unlike some former more "recreational" existences: at its inception, this *Physical Universe* was *created* as a place to penalize, exile, or otherwise imprison, the criminals and maladjusted individuals from the *Magic Universe.* This paints a much different picture of things than the fluffy politically-correct idea impressed by some *mystics* and *spiritual leaders* regarding this existence being some kind of "school." About the only thing an *Alpha-Spirit* has of value to *learn* "down here" is *why and how they got here* and *what is the best way out.*

Of course, there are some who *have* "escaped" in the past—and so there may be repeated *incidents* of *entry-into-this-Universe* (and as usual, one looks to *spot* the earliest one for *defragmentation*). But, this has no longer been the standard *goal* and *purpose* of a *Being* entrapped down here. For one: an individual no longer carries a clear *Knowingness* of their "entrapped" circumstances—so fewer attempts and efforts are even made.

There is another factor that has taken place as well: as more of the population from the *Magic Universe* ended up in the *Physical Universe*, there were fewer remaining to maintain the *"creation"* of the *Magic Universe*. Things eventually became *"more interesting"* in the *Physical Universe*, and many within the *Magic*

Universe migrated down here *knowingly* and by choice. Such individuals would at first have appeared to be quite *"advanced"* (or like *"gods"*) in contrast to the "more primitive" individuals entrapped in *Beta-Existence.*

The *transition-point* (involving the *"pool"*) is a significant experience to *"Spot"* when handling the *incident.* Unlike certain factors that are common to everyone that has the experience, the actual "reason" for such a "legal sentencing" is likely to be somewhat unique for each occurrence. [More data is still need in this area.]

For whatever the reason, an individual finds themselves being "drawn down" to descend a *"spiral of pillars"* into the *pool.* Whatever *Beings* are present (above/around the *pool*) also use their *energy* to "push down" upon the individual as well —thereby participating in the eventual *"implanting-incident"* that ensues. Once the transfer occurs, the *implanting-incid-*

ent, from this point onward, is always the same prerecorded pageantry.

THE IMPLANTING INCIDENT

Data for this *"implanting-incident"* accumulated over several years. It is consolidated from exploratory research from many *"advanced practitioners"* (working beyond *Beta-Defragmentation*) using *GSR-biofeedback devices* (and subsequently comparing notes afterward). As such, there are some details that may be in error (or absent), but enough valid information was recovered for a *Seeker* to *"Spot"* and *run* the *incident*. To start, let's simply look over the data.

This *implanting-incident* does not contain much "pain" (or "force"). It is a *"mini-verse"* manufactured in the *Magic Universe* to be *"aesthetically-beautiful"* so as to hold one's *attention* in and eventually draw the *Alpha-Spirit* into the *Physical Universe*. Any "*turbulence*" connected to the *incident* itself, is merely a sense of "*loss*" (over leaving the *Magic Universe*) or misemotion regarding "*exile*" (or being "*pushed out*").

There are really two parts to this *incident*: the *first* sets up the *second*. In the *first* part: an individual finds themselves essentially floating in a "*void-like space*" with a sensation similar to being "*under water.*" They eventually see a "*golden light*" in the distance—it glimmers and radiates like a sunburst, but more like a reflective piece of jewelry rather than an actual light-source itself.

Once the individual's *attention* is fixed upon it, the *implanting* begins. They are "drawn toward" it with increasing speed. This "*golden light*" is an "*Object-Item*" that is attached to the first "*Implanted Goal*" of an entire sequence (consisting of *36*

"*Goals*" or "*archetypes*"). [In this case: the first one, is "*To Be Godlike.*"]

Each "*Goal*" is introduced and identified with a particular "*Object-Item*" (in order to communicate and embed its significance). The "*Goals*" *are* named, but the titles are *sensed* (or *intuited*) rather than expressed in words; we can only *approximate* their full meaning with *Human* language (for *study* and *processing*).

The individual is "*drawn in and through*" each "*Object-Item*" (they are apparently holographic)—and while *passing through*, the "*Goal*" (and its label) is identified and understood. This continues with increasing speed through the remaining sequence—simply informing *what* the "*Goals*" *are*, in order to set up the *second* part of the *incident*, when this "*Goals-Sequence*" is *implanted* again with *significance.*

In the *second* part: the *Being* emerges from

the final *"Object-Item-Goal"* in this sequence (a *"pyramid"* associated with *"To Be Enduring"*) onto a different and more substantial plane or landscape that involves an "amphitheater" and "stage."

Some *Advanced Seekers* have referred to this whole *incident* as the *"Heaven Implant,"* because the scenery (of this *second* part) reflects a stereotypical angelic *"trumpet-blasting-cherubs-in-the-clouds"* atmosphere that *Humans* are *implanted* to "expect" for their *after life* (or more correctly: *"between-lives"* period). [In actual fact, the real *"Kingdom of Heaven"* (such as *Jesus* speaks of) is the *Magic Kingdom Universe*, and not this *implanting-incident* that everyone has been forced to experience (multiple times).]

The *second* part of the *incident* mainly entails a "theatrical pageant" or "skit." There is the appearance of other individuals in attendance; but it is unlikely that you would experience this simultaneously with other actual entities (*Alpha-Spirits*), so this crowd that gathers is probably fabricated as part of the recorded scene. Sometimes there is a feeling of being overwhelmed by the surrounding crowd as everyone excitedly pushes up toward the stage.

There is a trumpet-horn that blasts, and a sharp snapping-crackle (like the pierce of thunder), to get your *attention* right before any "*command-line*" is given. The first *command-line* is from an unseen *source* (to get the "pageant" started)—"*Only One Will Survive.*" Again, the horn blasts, the snaps happen, and the second *command-line* emerges—"*To Be The One Who Survives, You Must Be Superior To All Others.*" This seems to settle the crowd down and all *attention* becomes focused on the stage.

The blasts and snaps are also heard as a new "character" (each representing one of the "*Goals*") comes on the stage to speak *three* "command-lines." Only one "character" is on stage at a time. They always enter from one side of the stage (your right) and exit the other (your left). There is a *procession* or *sequence* taking place, so the "characters" tend to *look* in a particular direction when referring to the *next* (or the *previous*) "character."

When the *final* "character" (*An Enduring Being*) completes their reference to the *first* (*A Godlike Being*), the *implanting-incident* ends with "waves of blackness" before the individual finds themselves as an *Alpha-Spirit* with fixed *viewpoints* in *this Physical Universe* (*Beta-Existence*). And while there are many *entry-points* in this *local Universe*, according to research, the most recently used one is the *Horsehead Nebula* (*Orion*).

The *110* "*command-items*" of this *Implant-Platform* are given in the next section (below); and the significance for each is "*processed out*" (as described in *Lesson-11*

"*Spiritual Implants*"). The researched description (above) simply helps in "*Spotting*" and "*connecting with*" (*contacting*) each *Item* on the *Backtrack*. But before doing any *processing*, let's examine exactly what is taking place with this *Implant*.

The *incident* is *implanting* a *game*. The emphasis is "*survival*" rather than "*creation*." It is intended to sell you on the idea that the *game* of *this Universe* is "*survival by superiority*" — that you must be "*superior*" to everyone else, and only one will "*survive*" — which immediately puts you into *conflict* with everyone else and automatically sets you up for failure. And of course, there is no actual "*winning*" of the *Game*. It simply occupies *attention* while making its *Players* just a little bit worse off with each "*lifetime*."

When you're "fighting against" *everyone* else, it is only a matter of time before you "*lose*." While everyone is originally *implanted* to operate from a "*godlike*" state at the "start" of the *game*, no one is ever actually able to maintain that state (in opposition to everyone else);   But as we "fail" with each "*Goal*," our perceived "*purpose*" sinks down through the remaining sequence of "*Goals*"—until we end up at "*To Be Godlike*" again, but this time a "lesser version" of what we were before, as we continue down a dwindling spiral.

The entire "*Implant-Platform*" is a "lie"— built upon the foundation of the very first and most significant lie that "*only one will survive.*" This *implant* is the primary reason for conflict in this *Game-Universe*; the entire foundation of *this existence* is essentially "*bullshit.*"

When a *Player* finally calls "*bullshit*" on the *Game*, and can fully stop using their *skills* and *abilities* against others (and for *superiority*), then they can break free of the "*game-pattern*" that most strongly fixates *considerations* and *viewpoints* to *this*

Universe. Does *"processing-out"* the following *Implant-Platform* automatically release you from *Beta-Existence* and return you to the *Magic Universe?* Certainly not. But, if that *is* a direction you are intending to go, than this type of *implant-running* is a very critical step.

THE GOAL-SEQUENCING IMPLANT

The following *110 Command-Items* compose the basic *Implant-Platform* for *"Entry-to-this-Universe"* (and entry into *this Game*). Any personal significance attached to these *Items* is *processed-out* (using the training and procedure given in *Lesson-11*). While *processing*, try to *"Spot"* the *Items* *"in"* the *incident*, using the details described in the previous section (*imagining* when necessary).

0A. *"Only one will survive."*

0B. *"To be the one who survives, you must be superior to all others."*

1A. *"To be godlike is to solve the opposition of enduring (or stubborn) people."*

1B. *"To be godlike is to be superior to all others."*

1C. *"To be godlike is to suffer the oppression of free beings."*

2A. *"To be free is to solve the opposition of godlike beings."*

2B. *"To be free is to be superior to all others."*

2C. *"To be free is to suffer the oppression of responsible beings."*

3A. *"To be responsible is to solve the opposition of free beings."*

3B. *"To be responsible is to be superior to all others."*

3C. *"To be responsible is to suffer the oppression of creative beings."*

4A. *"To be creative is to solve the opposition of responsible beings."*

4B. *"To be creative is to be superior to all others."*

4C. *"To be creative is to suffer the oppression of important beings."*

5A. *"To be important is to solve the opposition of creative beings."*

5B. *"To be important is to be superior to all others."*

5C. *"To be important is to suffer the oppression of competent beings."*

6A. *"To be competent is to solve the opposition of important beings."*

6B. *"To be competent is to be superior to all others."*

6C. *"To be competent is to suffer the oppression of famous beings."*

7A. *"To be famous is to solve the opposition of competent beings."*

7B. *"To be famous is to be superior to all others."*

7C. *"To be famous is to suffer the oppression of perceptive beings."*

8A. *"To be perceptive is to solve the opposition of famous beings."*

8B. *"To be perceptive is to be superior to all others."*

8C. *"To be perceptive is to suffer the oppression of energetic beings."*

9A. *"To be energetic is to solve the opposition of perceptive beings."*

9B. *"To be energetic is to be superior to all others."*

9C. *"To be energetic is to suffer the oppression of meticulous beings."*

10A. *"To be meticulous is to solve the opposition of energetic beings."*

10B. *"To be meticulous is to be superior to all others."*

10C. *"To be meticulous is to suffer the oppression of successful beings."*

11A. *"To be successful is to solve the opposition of meticulous beings."*

11B. *"To be successful is to be superior to all others."*

11C. *"To be successful is to suffer the oppression of accurate beings."*

12A. *"To be right (accurate) is to solve the opposition of successful beings."*

12B. *"To be right (accurate) is to be superior to all others."*

12C. *"To be right (accurate) is to suffer the oppression of popular beings."*

13A. *"To be popular is to solve the opposition of accurate beings."*

13B. *"To be popular is to be superior to all others."*

13C. *"To be popular is to suffer the oppression of skillful beings."*

14A. *"To be skillful is to solve the opposition of popular beings."*

14B. *"To be skillful is to be superior to all others."*

14C. *"To be skillful is to suffer the oppression of wise beings."*

15A. *"To be wise is to solve the opposition of skillful beings."*

15B. *"To be wise is to be superior to all others."*

15C. *"To be wise is to suffer the oppression of beautiful beings."*

16A. *"To be beautiful is to solve the opposition of wise beings."*

16B. *"To be beautiful is to be superior to all others."*

16C. *"To be beautiful is to suffer the oppression of productive beings."*

17A. *"To be productive is to solve the opposition of beautiful beings."*

17B. *"To be productive is to be superior to all others."*

17C. *"To be productive is to suffer the oppression of powerful beings."*

18A. *"To be powerful is to solve the opposition of productive beings."*

18B. *"To be powerful is to be superior to all others."*

18C. *"To be powerful is to suffer the oppression of holy beings."*

19A. *"To be holy is to solve the opposition of powerful beings."*

19B. *"To be holy is to be superior to all others."*

19C. *"To be holy is to suffer the oppression of intellectual beings."*

20A. *"To be intelligent is to solve the opposition of holy beings."*

20B. *"To be intelligent is to be superior to all others."*

20C. *"To be intelligent is to suffer the oppression of strong beings."*

21A. *"To be strong is to solve the opposition of intellectual beings."*

21B. *"To be strong is to be superior to all others."*

21C. *"To be strong is to suffer the oppression of crafty beings."*

22A. *"To be crafty is to solve the opposition of strong beings."*

22B. *"To be crafty is to be superior to all others."*

22C. *"To be crafty is to suffer the oppression of brave beings."*

23A. *"To be brave is to solve the opposition of crafty beings."*

23B. *"To be brave is to be superior to all others."*

23C. *"To be brave is to suffer the oppression of wealthy beings."*

24A. *"To be wealthy is to solve the opposition of brave beings."*

24B. *"To be wealthy is to be superior to all others."*

24C. *"To be wealthy is to suffer the oppression of independent beings."*

25A. *"To be independent is to solve the opposition of wealthy beings."*

25B. *"To be independent is to be superior to all others."*

25C. *"To be independent is to suffer the oppression of morally good beings."*

26A. *"To be good is to solve the opposition of independent beings."*

26B. *"To be good is to be superior to all others."*

26C. *"To be good is to suffer the oppression of adventurous beings."*

27A. *"To be adventurous is to solve the opposition of good beings."*

27B. *"To be adventurous is to be superior to all others."*

27C. *"To be adventurous is to suffer the oppression of orderly (organized) beings."*

28A. *"To be orderly is to solve the opposition of adventurous beings."*

28B. *"To be orderly is to be superior to all others."*

28C. *"To be orderly is to suffer the oppression of different (eccentric) beings."*

29A. *"To be different is to solve the opposition of orderly (organized) beings."*

29B. *"To be different is to be superior to all others."*

29C. *"To be different is to suffer the oppression of respected beings."*

30A. *"To be respected is to solve the opposition of different (eccentric) beings."*

30B. *"To be respected is to be superior to all others."*

30C. *"To be respected is to suffer the oppression of happy beings."*

31A. *"To be happy is to solve the opposition of respected beings."*

31B. *"To be happy is to be superior to all others."*

31C. *"To be happy is to suffer the oppression of acquisitive beings."*

32A. *"To be acquisitive is to solve the opposition of respected beings."*

32B. *"To be acquisitive is to be superior to all others."*

32C. *"To be acquisitive is to suffer the oppression of sensual beings."*

33A. *"To be sensual is to solve the opposition of acquisitive beings."*

33B. *"To be sensual is to be superior to all others."*

33C. *"To be sensual is to suffer the oppression of domineering beings."*

34A. *"To be domineering is to solve the opposition of sensual beings."*

34B. *"To be domineering is to be superior to all others."*

34C. *"To be domineering is to suffer the oppression of tough beings."*

35A. *"To be tough is to solve the opposition of domineering beings."*

35B. *"To be tough is to be superior to all others."*

35C. *"To be tough is to suffer the oppression of enduring (or stubborn) beings."*

36A. *"To be enduring is to solve the opposition of domineering beings."*

36B. *"To be enduring is to be superior to all others."*

36C. *"To be enduring is to suffer the oppression of godlike beings."*

GAMES AND UNIVERSES

The subject area of *"Games and Universes"* is quite advanced, even for our *Systemology*. In essence, a *Seeker* is *indirectly* treating this same area with *systematic processing* the whole while they are progressing on the *Pathway*—right from the beginning of *Systemology Level-0*. But, when it comes to *really understanding* what is happening with us and all around us, it essentially boils down to *"Games and Universes."*

It should be evident from this lesson that *this present Game-Universe* is based on *"force"* and *"conflict"*—which are used toward achieving a *"superiority"* that never actually arrives. This is not so surprising, given the original *intention* for this *Beta-Existence* as primarily a *"spirit-prison."* But, this is not the first *Game-Universe* that developed as a "penalty" of sorts. Prior to this, the *Magic Universe* also once began as a *secondary universe* to the one that preceded it. So these "cycles" have been taking place for a very long time.

Game-Universes aren't "new"; they actually start to appear quite early on the *Backtrack*. And *Alpha-Spirits* as *"Eternal Beings"* really like to play *"games"*—to *have* something to *do* in order to *Be*. This is the reverse of what our original native state is, where we can simply *Be* anything —and even *create* anything—*at will*.

But, as is even evident by the popularity of certain types of *"video games"* today, the *Alpha-Spirit* likes to *"play"* when it is not engaged in *"creating"* things to play with. And, of course, this has gotten us into a bit of trouble along the long course of our *Spiritual Existence*. The *Alpha-Spirit* eventually lost the *willingness* and *ability* to *confront* the *Infinity-of-Nothingness*, and now prefers literally *any game* over that native state.

In the earliest *Shared-Universes*, *games* were handled with much more primitive novelty—not very different from modern "*motion pictures*," "*video games*," and even "*virtual reality*." They were conducted *knowingly* for *entertainment*; a way to interactively "show off" our *creations* to one another. They were never meant to reach such a level as to *unknowingly entrap spirits* or employ *actual living things to suffer as playing pieces*—and yet this is the real truth about how far we have descended in our existence as *Spiritual Beings*.

In order to progress beyond the upper-most reaches of the *Pathway*, a *Seeker* will have to eventually regain the use of *force* and *energy* (from a perspective as an *Alpha-Spirit*, not a *Human*). This area of development is only emphasized *after* a *Seeker* can be certain that they are master-ing it to regain *control* of the *conditions* that have *entrapped* them, and not to use it to further the *domination/superiority-game* with others. [Doing so is technically how we *lost* or *forgot* these *spiritual abilities* in the first place! If you *knowingly* continue that *game* after increasing *Awareness* and personal power on this course, you will likely sink so fast, like in quicksand, that even *we* might not be able to pull you back out again.]

There is nothing inherently wrong with *games*. There is nothing even inherently wrong with the *"Goals"* of *this Universe*, themselves. Our own goal in *defragment-ing "The Goals"* is for a *Seeker* to retain the "positive" characteristics that each sug-gests, without being compelled to adopt the "inverted" *considerations* that they should be "weaponized" or used against others (which is what the *Implant* is really *impressing*).

By following along with the *Implant* (*Sequence*) we encounter unnecessary invisible *barriers* that restrict our very *considerations* of *Beingness*. For example: the *Implant* suggests that we should have issues *being* both *strong* and *intelligent*, or with *being responsible* and a *Free Spirit*, *&tc.* While the "*command-items*" do not literally say these things, these are the type of *Alpha-Thoughts* ("*postulates*") that an individual will *consider* on their own for themselves purely as a result of having experienced the *Implant*.

This whole area of "*Implanted-Goals*" is very closely associated with "*Justification Considerations*" (see *Lesson-8*) that *Seekers* uncovered about themselves previously to complete *Systemology Level-3*. This is where we did a bit of hunting to find out "*what makes one superior*" and "*what makes others wrong."* The answers to this will often provide a clue as to where on the *"Goal-Sequencing cycle"* a *Seeker* is likely to be. In other words: *Justification Considerations* usually indicate which *Implanted-Goal* a *Seeker* is presently occupied with.

One exercise that can help free up these *considerations* is: walk around and *get the sense* (or *Imagine*) yourself as having the primary positive characteristic of a *"Goal."* Then *look* around and *Spot* different people, *Imagining* that each of them also carries that strong positive characteristic (even if they are not visibly demonstrating it). Now, *Imagine* that there is *more* of that characteristic—being *more godlike*, or *stronger*, *&tc.* Alternate: putting the *intention* into yourself, and putting the *intention* into others, [Keep in mind that we have been "competitively programmed" to actually maintain the opposite *intentions* about others.]

There are some *advanced processes* concerning *Universes* that are used as a "checkpoint" for *Seekers*, here at the completion of *Systemology Level-5*. They may be used to increase "*spiritual perception*," but for our present purposes, they are intended to "*unstick*" or "*unfix*" a *Seeker's attention* and *Awareness* from being so compulsively on *this Game-Universe* (*Beta-Existence*).

The first exercise is adapted from the original "*Wizard Training Regimen*" developed by the *Systemology Society* many years ago. The second/final *process* is an advanced application of a familiar technique we've used previously in the *Professional Course.* These are "*creativeness processing*" exercises that may be *run* using one's "*imagination*" (or "*visualization*") until a *Seeker* gets a sense that their *spiritual perception* has increased.

* EXERCISE #1 *

Sitting comfortably, eyes closed.

• *Imagine* you are extending your *Awareness*, reaching through the entire *Physical Universe* (as far as you can imagine). Now *reach* a bit further beyond all perception of dimensional space until you find "*Nothing.*" Hold your *attention* (*point-of-view*) on the "*Nothingness*" without thinking of, or imagining, anything else.

• As before: specifically extend your reach out on the *right side*; contact and hold on the "*Nothingness.*"

• Practice as before: extending your reach out to each direction (by itself); the *left side*, *in-front*, *behind*, *above*, and *below*.

• Now extend your reach out in two directions (simultaneously); making sure to contact and get a sense of the "*Nothingness*" on both sides (from your *point-of-view*)—*right and left*; then *in-front* and *behind*; and finally, *above* and *below*.

• When a *Seeker* is well practiced in the above: extend your reach in all six directions simultaneously, getting a full certainty of the "*Nothingness*" on all sides.

* EXERCISE #2 *

Laying down, comfortably, eyes closed.

Alternate.

A. "*Spot three points in this Universe.*"

B. "*Spot three points that are not in this Universe.*"

This completes *Systemology Level-5*.

The Systemology Professional Course
continues in the next lesson booklet:
SPIRITUAL ENERGY

GLOSSARY

actualization : to make actual, not just potential; to bring into full solid Reality; to realize fully in *Awareness* as a "thing."

agreement (reality) : unanimity of opinion of what is "thought" to be known; an accepted arrangement of how things are; things we consider as "real" or as an "is" of "reality"; a consensus of what is real as made by standard-issue (common) participants; what an individual contributes to or accepts as "real"; in *Systemology*, a synonym for "*reality.*"

alpha : the first, primary, basic, superior or beginning of some form; in *Systemology*, referring to the state of existence operating on spiritual archetypes and postulates, will and intention "exterior" to the low-level condensation and solidarity of energy and matter as the 'physical universe' (*beta*).

alpha-spirit : a "spiritual" *Life*-form; the "true" *Self* or I-AM; the *individual*; the spiritual (*alpha*) *Self* that is animating the (*beta*) physical body or "*genetic vehicle*" using a continuous *Lifeline* of spiritual ("*ZU*") energy; an individual spiritual (*alpha*) entity possessing no physical mass or measurable waveform (motion) in the Physical Universe as itself, so it animates the (*beta*) physical body or "*genetic vehicle*" as a catalyst to experience *Self*-determined causality in effect within the *Physical Universe*; a singular unit or point of *Spiritual Awareness* that is *Aware* that it is *Aware*.

alpha thought : the highest spiritual *Self-determination* over creation and existence exercised by an Alpha-Spirit; the Alpha range of pure *Creative Ability* based on direct postulates and considerations of *Beingness*; spiritual qualities comparable to "thought" but originating in Alpha-existence, independently superior to a Mind-System.

ascension : actualized *Awareness* elevated to the point of true "spiritual existence" exterior to *beta existence*. An "Ascended Master" is one who has returned to an incarnation on Earth as an inherently *Enlightened One*, demonstrable in their words and actions; they have the ability to *Self-direct* the "Mind" and "Body" as *Self* (as a "Spirit"); and to maintain consciousness as a personal identity continuum with the same *Self-directed* control and communication of Will-Intention that is exercised, actualized and developed deliberately during one's present incarnation.

associative knowledge : significance or meaning of a facet or aspect assigned to (or considered to have) a direct relationship with another facet; to connect or relate ideas or facets of existence with one another; in traditional systems logic, an equivalency of significance or meaning between facets or sets that are grouped together, such as in $(a + b) + c = a + (b + c)$; in Systemology, erroneous associative knowledge is assignment of the same value to all facets or parts considered as related (even when they are not actually so), such as in $a = a, b = a, c = a$ and so forth without distinction.

attention : active use of *Awareness* toward a specific aspect or thing; the act of "attending" with the presence of *Self*; a direction of focus or concentration of *Awareness* along a particular channel or conduit or toward a particular terminal node or communication termination point; the Self-directed concentration of personal energy as a combination of observation, thought-waves and consideration; focused application of *Self-Directed Awareness*.

awareness : the highest sense of-and-as *Self* in knowing and being as I-AM (the *Alpha-Spirit*); the extent of beingness directed as a viewpoint (POV) experienced by *Self* as *Knowingness*.

beta (awareness) : all consciousness activity ("*Awareness*") in the "Physical Universe" (KI, in *Zuism*) or else in *beta-existence*; *Awareness* within the range of the *genetic-body*, including material thoughts, emotional responses and physical motors; personal *Awareness* of physical energy and physical matter moving through physical space and experienced as "time"; the *Awareness* held by *Self* that is restricted to an organic *Lifeform* or "*genetic vehicle*" in which it experiences causality in *beta-existence*.

beta (existence) : all manifestation in the "Physical Universe" (KI, in *Zuism*); the conditions of *Awareness* for the *Alpha-spirit* (*Self*) as a physical organic *Lifeform* or "*genetic vehicle*" in which it experiences causality in the *Physical Universe*.

charge : to fill or furnish with a quality; to supply with energy; to lay a command upon; in *Systemology*—to imbue with intention; to overspread with emotion; personal energy stores and significances entwined as fragmentation in mental images, reactive-response encoding and intellectual (and/or) programmed beliefs.

channel : a specific stream, course, current, direction or route; to form or cut a groove or ridge or otherwise guide along a specific course; a direct path; an artificial aqueduct created to connect two water bodies or water or make travel possible.

circuit : a circular path or loop; a closed-path within a system that allows a flow; a pattern or action or wave movement that follows a specific route or potential path only; in *Systemology*, "*communication processing*" pertaining to a specific *flow* of energy or information along a channel; "*feedback loop.*"

communication : successful transmission of information, data, energy (&tc.) along a message line, with a reception of feedback; an energetic flow of intention to cause an effect (or duplication) at a distance; the personal energy moved or acted upon by will or else 'selective directed attention'; the 'messenger action' used to transmit and receive energy across a medium; also relay of energy, a message or signal—or even locating a personal POV (viewpoint) for the Self—along the *ZU-line*.

condense (condensation) : the transition of vapor to liquid; denoting a change in state to a more substantial or solid condition; leading to a more compact or solid form.

confront : to come around in front of; to be in the presence of; to stand in front of, or in the face of; to meet "face-to-face" or "face-up-to"; additionally, in *Systemology*, to fully tolerate or acceptably withstand an encounter with a particular manifestation without an automatic reactive response.

consideration : careful analytical reflection of all aspects; deliberation; determining the significance of a "thing" in relation to similarity or dissimilarity to other "things"; evaluation of facts and importance of certain facts; thorough examination of all aspects related to, or important for, making a decision; the analysis of consequences and estimation of significance when making decisions; also in *Systemology*, the *postulate* or *Alpha-Thought* that defines the state of *beingness* for what something "*is.*"

defragmentation : the *reparation* of wholeness; collecting all dispersed parts to reform an original whole; a process of removing "*fragmentation*" in data or knowledge to provide a clear understanding; applying techniques and processes that promote a *holistic* interconnected *alpha* state, favoring observational *Awareness* of continuity in all spiritual and physical systems; in *Systemology*, a "*Seeker*" achieving actualized "*Self-Honest Awareness*" is said to be in a basic state of *beta-defragmentation*, whereas *Alpha-defragmentation* is the rehabilitation of the *creative ability*, managing the *Spiritual Timeline* and the POV of *Self* as Alpha-Spirit (I-AM).

existence : the *state* or fact of *apparent manifestation*; the resulting combination of the Principles of Manifestation: consciousness, motion and substance; continued *survival*; that which independently exists.

exterior : outside of; on the outside; in *Systemology*, we mean specifically the POV of *Self* that is *'outside of'* the *Human Condition,* free of the physical and mental trappings of the Physical Universe; a metahuman range of consideration; see also '*Zu-Vision*'.

external : a force coming from outside; information received from outside sources; in *Systemology*, the objective *'Physical Universe'* existence, or *beta-existence*, that the Physical Body or *genetic vehicle* is essentially *anchored* to for its considerations of locational space-time as a dimension or POV.

fragmentation : breaking into parts and scattering the pieces; the *fractioning* of wholeness or the *fracture* of a holistic interconnected *alpha* state, favoring observational *Awareness* of perceived connectivity between parts; *discontinuity*; separation of a totality into parts; in *Systemology*, a person outside of *Self-Honesty* is said to be operating from a *fragmented* state.

flow : movement across (or through) a channel (or conduit); a direction of active energetic motion, typically distinguished as either an *in-flow*, *out-flow* or *cross-flow*.

genetic-vehicle : a physical *Life*-form; the physical (*beta*) body that is animated/controlled by the (*Alpha*) *Spirit* using a continuous *Spiritual Lifeline* (ZU); a physical (*beta*) organic receptacle and catalyst for the (*Alpha*) *Self* to operate "causes" and experience "effects" within the *Physical Universe*.

harmful-act : a counter-survival mode of behavior or action (esp. that causes harm to one of more *Spheres of Existence*)—or—an overtly aggressive (hostile and/or destructive) action against an individual or any other *Sphere of Existence*; in *Utilitarian Systemology*—a shortsighted (serves fewest/lowest *Spheres of Existence*) intentional overtly harmful action to resolve a perceived problem; a revision of the rule for standard *Utilitarianism* for Systemology to distinguish actions which provide the least benefit to the least number of *Spheres of Existence*, or else the greatest harm to the greatest number of *Spheres of Existence*; in *moral philosophy*—an action which can be experienced by few and/or which one would not be willing to experience for themselves (*theft, slander, rape, &tc*); an iniquity or iniquitous act.

hold-back : withheld communications (esp. actions) such as "*Hold-Outs*"; intentional (or automatic) withdrawal (as opposed to reach); Self-restraint (which may eventually be enforced or automated); not reaching, acting or expressing, when one should be; an ability that is now restrained (on automatic) due to inability to withhold it on Self-determinism alone.

hold-outs : in photography, the numerous snapshots/pictures withheld from the final display or professional presentation of the event; withheld communications; in Utilitarian Systemology—energetic withdrawal and communication breaks with a "*terminal*" and its *Sphere of Existence* as a result of a "*Harmful-Act*"; unspoken or undiscovered (hidden, covert) actions that an individual withholds communications of, fearing punishment or endangerment of *Self-preservation* (*First Sphere*); the act of hiding (or keeping hidden) the truth of a "*Harmful-Act*"; a refusal to communicate with a *Pilot*; also "*Hold-Back.*"

holistic : the examination of interconnected systems as encompassing something greater than the *sum* of their "parts."

Human Condition : a standard default state of Human experience that is generally accepted to be the extent of its potential identity (*beingness*) —currently treated as *Homo Sapiens Sapiens,* but which is scheduled for replacement by *Homo Novus* (the "New Human").

imagination : ability to create *mental imagery* in one's Personal Universe at will and change or alter it as desired; the ability to create, change and dissolve mental images on command or as an act of will; to create a mental image or have associated imagery displayed (or "conjured") in the mind that may or may not be treated as real (or memory recall) and may or may not accurately duplicate objective reality; to employ *creative abilities* of the Spirit that are independent of reality agreements with beta-existence.

imprint : to strongly impress, stamp, mark (or outline) onto a softer 'impressible' substance; to mark with pressure onto a surface; in *Systemology*, used to indicate permanent Reality impressions marked by frequencies, energies or interactions experienced during periods of emotional distress, pain, unconsciousness, loss, enforcement, or something antagonistic to physical (personal) survival, all of which are are stored with other reactive response-mechanisms at lower-levels of *Awareness* as opposed to the active memory database and proactive processing center of the Mind; an experiential "memory-set" that may later resurface—be triggered or stimulated artificially—as Reality, of which similar responses will be engaged automatically; holographic-like imagery "stamped" onto consciousness as composed of energetic *facets* tied to the "snap-shot" of an experience.

imprinting incident : the first or original event

instance communicated and *emotionally encoded* onto an individual's "*Spiritual Timeline*" (recorded memory from all lifetimes), which formed a permanent impression that is later used to mechanistically treat future contact on that channel; the first or original occurrence of some particular *facet* or mental image related to a certain type of *encoded response*, such as pain and discomfort, losses and victimization, and even the acts that we have taken against others along the *Spiritual Timeline* of our existence that caused them to also be *Imprinted*.

intention : directed application of Will; to intend (have "in Mind") or signify (give "significance" to) for or toward a particular purpose; in *Systemology* (from the *Standard Model*)—the spiritual activity at WILL (5.0) directed by an *Alpha Spirit* (7.0); the application of WILL as "Cause" from a higher order of Alpha Thought and consideration (6.0).

interior : inside of; on the inside; in *Systemology*, we mean specifically the POV of *Self* that is fixed to the *'internal' Human Condition,* including the *Reactive Control Center* (RCC) and Mind-System or *Master Control Center* (MCC); within *beta-existence*.

internal : a force coming from inside; information received from inside sources; in *Systemology*, the objective experience of *beta-existence* associated with the Physical Body or *genetic vehicle* and its POV regarding sensation and perception; from inside the body; in the body.

invalidate : decrease the level or degree or *agreement* as Reality.

mental image : a subjectively experienced "picture" created and imagined into being by the Alpha-Spirit (or at lower levels, one of its automated mechanisms) that includes all perceptible *facets* of totally immersive scene, which may be forms originated by an individual, or a "facsimile-copy" ("snap-shot") of something seen or encountered; a duplication of wave-forms in one's Personal Universe as a "picture" that mirror an "external" Universe experience, such as an *Imprint*.

perception : internalized processing of data received by the *senses*; to become *Aware of* via the senses.

pilot : a professional steersman responsible for healthy functional operation of a ship toward a specific destination; in *Systemology*, an intensive trained individual qualified to specially apply *Systemology Processing* to assist other *Seekers* on the *Pathway*.

point-of-view (POV) : a point to view from; an opinion or attitude as expressed from a specific identity-phase; a specific standpoint or vantage-point; a definitive manner of consideration specific to an individual phase or identity; a place or position affording a specific view or vantage; circumstances and programming of an individual that is conducive to a particular response, consideration or belief-set (paradigm); a position (consideration) or place (location) that provides a specific view or perspective (subjective) on experience (of the objective).

postulate : to put forward as truth; to suggest or assume an existence *to be*; to state or affirm the existence of particular conditions; to provide a basis of reasoning and belief; a basic theory accepted as fact; in *Systemology*, Alpha-Thought —the top-most decisions or considerations made by the Alpha-Spirit regarding the "*isness*" (what things "are") about energy-matter and space-time.

presence : a quality of some thing (*energy/matter*) being "present" in space-time; personal orientation of *Self* as an *Awareness* (*POV*) located in present space-time (environment) and communicating with extant energy-matter.

processing command line (PCL) : a directed input; a specific command using highly selective language for *Systemology Processing*; a predetermined directive statement (cause) intended to focus concentrated attention (effect).

processing, systematic : the inner-workings or "through-put" result of systems; in *Systemology*, a method of applied spiritual technology used toward personal Self-Actualization; methods of selective directed attention, communicated language and associative imagery that increases personal control of the human condition.

realization : the clear perception of an understanding; a consideration or understanding on what is "actual"; to make "real" or give "reality" to so as to grant a property of "being-ness" or "being as it is"; the state or instance of coming to an *Awareness*; in *Systemology*, "gnosis" or true knowledge achieved during *systematic processing*; achievement of a new (or higher) cognition, true knowledge or perception of Self; a consideration of reality or assignment of meaning.

responsibility : the *ability* to *respond*; the extent of mobilizing *power* and *understanding* an individual maintains as *Awareness* to enact *change*; the proactive ability to *Self-direct* and make decisions independent of an outside authority.

Seeker : an individual on the *Pathway to Self-Honesty*; a practitioner of *Mardukite Systemology* or *Systemology Processing*, that is working toward *Spiritual Ascension*.

Self-actualization : bringing the full potential of the Human spirit into Reality; expressing full capabilities and creativeness of the *Alpha-Spirit*.

Self-determinism : the freedom to act, clear of external control or influence; the personal control of Will to direct intention.

Self-honesty : the basic or original *alpha* state of *being* and *knowing*; clear and present total *Awareness* of-and-as *Self*, in its most basic and true proactive expression of itself as *Spirit* or *I-AM*—free of artificial attachments, perceptive filters and other emotionally-reactive or mentally-conditioned programming imposed on the human condition by the systematized physical world; the ability to experience existence without judgment.

spiritual timeline : a continuous stream of moment-to-moment *Mental Images* (or a record of experiences) that defines the "past" of a spiritual being (or *Alpha-Spirit*) and which includes impressions (*imprints, &tc.*) from all life-incarnations and significant spiritual events the being has encountered; in Systemology, also "*backtrack*."

Spheres of Existence : a series of *eight* concentric circles, rings or spheres (each larger than the former) that is overlaid onto the Standard Model of Beta-Existence to demonstrate the dynamic systems of existence extending out from the POV of Self (often as a "body") at the *First Sphere*; these are given in the basic eightfold systems as: *Self, Home/Family, Groups, Humanity, Life on Earth, Physical Universe, Spiritual Universe* and *Infinity-Divinity.*

Systemology : a modern tradition of applied religious philosophy and spiritual technology based on *Arcane Tablets* (in combination with "*general systemology*" and "*games theory*") developed in the New Age underground by Joshua Free in 2011 as an advanced futurist extension of the *Mardukite Research Org.*

terminal (node) : a point, end, or mass, on a line; a connection point for closing an electric circuit, such as a post on a battery terminating at each end of its own systematic function; a point of connectivity with other points; in systems, a contact point of interaction; a point of interaction with other points.

turbulence : a quality or state of distortion or disturbance that creates irregularity of a flow or pattern; the quality or state of aberration on a line (such as ragged edges) or the emotional "turbulent feelings" attached to a particular flow or terminal node; a violent, haphazard or disharmonious commotion (such as in the ebb of gusts and lulls of wind action).

validation : a reinforcement of agreements or considerations as being "real."

viewpoint : see "*point-of-view*" *(POV)*.

willingness : the state of conscious Self-determined ability and interest (directed attention) to *Be*, *Do* or *Have*; a Self-determined consideration to reach, face up to (*confront*) or manage some "mass" or energy; the extent to which an individual considers themselves able to participate, act or communicate along some line, to put attention or intention on the line, or to produce (create) an effect.

ZU : the ancient Sumerian cuneiform sign for the archaic verb—"*to know*," "*knowingness*" or "*awareness*"; in *Mardukite Zuism and Systemology*, the active energy/matter of the "Spiritual Universe" (AN) experienced as a *Lifeforce* or *consciousness* that imbues living forms extant in the "Physical Universe" (KI); "*Spiritual Life Energy*"; energy demonstrated by the WILL of an actualized *Alpha-Spirit* in the "Spiritual Universe" (AN), which impinges its *Awareness* into the Physical Universe (KI), animating/controlling *Life* for its experience of *beta-existence* along an individual Alpha-Spirit's personal *Identity-continuum*, called a *ZU-line*.

*Zu-*Line : a theoretical construct in *Mardukite Zuism and Systemology* demonstrating *Spiritual*

Life Energy (*ZU*) as a personal individual "continuum" of Awareness interacting with all Spheres of Existence on the Standard Model of Systemology; a spectrum of potential variations and interactions of a monistic continuum or singular *Spiritual Life Energy* demonstrated on the Standard Model; an energetic channel of potential POV and "locations" of Beingness, demonstrated in early Systemology materials as an individual Alpha-Spirit's personal *Identity- continuum*, potentially connecting *Awareness* of *Self* with "*Infinity*" simultaneous with all points considered in existence; a symbolic demonstration of the "*Life-line*" on which *Awareness (ZU)* extends from the direction of the "Spiritual Universe" (AN) in its true original *alpha state* through an entire possible range of activity resulting in its *beta state* and control of a *genetic-entity* occupying the *Physical Universe (KI).*

Zu-**Vision** : the true and basic (*Alpha*) Point-of-View (perspective, POV) maintained by *Self* as *Alpha-Spirit* outside boundaries or considerations of the *Human Condition* and *exterior* to beta-existence reality agreements with the Physical Universe; a POV of Self *as* "a unit of Spiritual Awareness" that exists independent of a "body" and entrapment in a *Human Condition*; "spirit vision" in its truest sense.

explore the
Fundamentals of Systemology

All *six*
Basic Course
lesson booklets
in one
hardcover
edition!

start your journey on the
The Pathway to Ascension

All *sixteen*
Professional Course
lesson booklets
in two
hardcover
volumes!

THE SYSTEMOL

Seekers and students of the *Basic Course* and *Professional Course* will also be interested in the *Systemology Core Research Series*. These eight volumes are a complete chronological record of the Mardukite New Thought developments from the Systemology Society, published in 2019 through 2023.

The *Systemology Core* begins with the first professional publication released when the *Mardukite Systemology Society* emerged from the underground in 2019, with: *"The Tablets of Destiny Revelation."*

OGY PATHWAY

The Tablets of Destiny Revelation:
*How Long-Lost Anunnaki Wisdom
Can Change the Fate of Humanity*

Crystal Clear: *Handbook for Seekers*

Metahuman Destinations (2 *volumes*)

Imaginomicon:
Approaching Gateways to Higher Universes

Way of the Wizard: *Utilitarian Systemology*

Systemology-180: *Fast-Track to Ascension*

Systemology Backtrack:
Reclaiming Spiritual Power & Past-Life Memory

PUBLISHED BY THE **JOSHUA FREE** IMPRINT REPRESENTING

The Mardukite Academy of Systemology

mardukite.com

www.ingramcontent.com/pod-product-compliance
Lightning Source LLC
Chambersburg PA
CBHW071208120626
46546CB00006B/2467